Beautiful Birds

CARDINALS

By Bray Jacobson

Gareth Stevens
PUBLISHING

Please visit our website, www.garethstevens.com. For a free color catalog of all our high-quality books, call toll free 1-800-542-2595 or fax 1-877-542-2596.

Library of Congress Cataloging-in-Publication Data

Names: Jacobson, Bray, author.
Title: Cardinals / Bray Jacobson.
Description: Buffalo, New York : Gareth Stevens Publishing, [2024] | Series: Beautiful birds | Includes index. | Audience: Grades K-1
Identifiers: LCCN 2023023589 (print) | LCCN 2023023590 (ebook) | ISBN 9781538292440 (library binding) | ISBN 9781538292433 (paperback) | ISBN 9781538292457 (ebook)
Subjects: LCSH: Cardinals (Birds)–Juvenile literature.
Classification: LCC QL696.P2438 J33 2024 (print) | LCC QL696.P2438 (ebook) | DDC 598.8/83–dc23
LC record available at https://lccn.loc.gov/2023023589
LC ebook record available at https://lccn.loc.gov/2023023590

First Edition

Published in 2024 by
Gareth Stevens Publishing
2544 Clinton St.
Buffalo, NY 14224

Editor: Kristen Nelson
Designer: Andrea Davison-Bartolotta

Photo credits: Cover, pp. 1, 9, 24 (bill) Bonnie Taylor Barry/Shutterstock.com; p. 5 Andrea J Smith/Shutterstock.com; pp. 7, 24 (crest) DavidM326/Shutterstock.com; pp. 11, 24 (seeds) Steve Collender/Shutterstock.com; p. 13 Barb Elkin/Shutterstock.com; p. 15 (main) Mike Laptev/Shutterstock.com; p. 15 (inset) April Hayes/Shutterstock.com; p. 17 Chris Klonowski/Shutterstock.com; p. 19 Alan B. Schroeder/Shutterstock.com; p. 21 Cathleen Wake Gorbatenko/Shutterstock.com; p. 23 WoodysPhotos/Shutterstock.com.

Printed in the United States of America

CPSIA compliance information: Batch #CW24GS: For further information contact Gareth Stevens, New York, New York at 1-800-542-2595.

Contents

I see a red bird!
It is a male cardinal.

It has feathers
on its head.
This is a crest.

It has a long tail.
It has a pointy
orange bill.

It eats bugs and berries.
It likes seeds too.

It lives in forests.
It hides in trees.

Nests are made
of twigs and bark.
They use grass too.

Mothers are mostly brown. They lay three to four eggs.

Eggs can be white, blue, or green. They can have brown spots!

Mothers sit
on the eggs.
They sing while they sit!

Parents feed the babies.

Words to Know

bill

crest

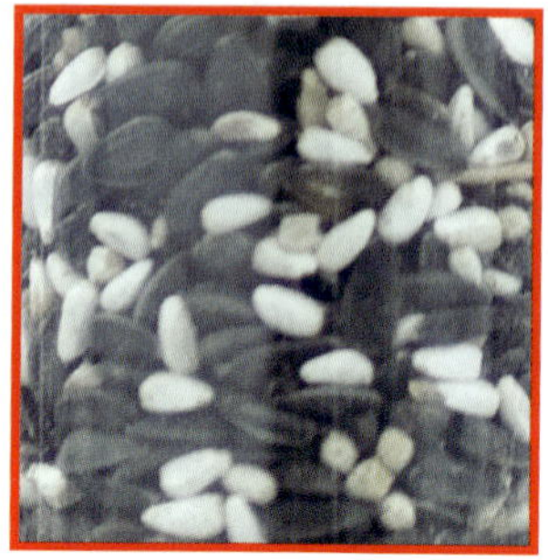

seeds

Index